HOW THE UNIVERSE WORKS

Albert W. McKinney III
2018 June 7

DEDICATION

This book is dedicated to Phyllis McKinney, my wife of 67 years.

Library of Congress Cataloging-in-Publication Data
McKinney, Albert William III (1929-)
     How the Universe Works
     CreateSpace Independent Publishing Platform,
         North Charleston, South Carolina
     ISBN:9781720916369

TABLE OF CONTENTS

Preface

This book is not an attack on ideas. Ideas, right or wrong, are often necessary steps toward advances in physics and astronomy.

But wrong ideas need to be identified and either corrected or scrapped. And those objectives are two of the reasons for this book.

The scrapped ideas are all from the twentieth century.

My criterion for scrapping an idea is whether or not it is consistent with the theory of connections, which is presented at the end of this book. Some ideas of that theory are also tucked into corners here and there earlier in the book.

Connection theory shows how the basic ideas of physics (such as electromagnetism, the strong nuclear force, gravity, light, as well as the precession of the perihelia of planets orbiting the Sun) can be explained without photons, gravitons, quarks, fields, general relativity, string theory, etc.

This book is intended as a *description* of the physical properties at work in the universe. Some of the mathematical and physical statements made herein are not explicitly proven in this book, although the proofs are sometimes suggested. In general, more explicit proofs will be found in the the references.

The description covers what exists in our part of the universe, excluding what happens in laboratories equipped with massive physical machines. I do not deny the existence of particles created there, but I do not see how to fit them into the universe.

Those who wish to comment on the ideas of this book may do so by e-mail. My e-mail address is mickeymck@prodigy.net.

PART 1: SOME IDEAS OF PHYSICS AND ASTRONOMY CORRECTED OR SCRAPPED

CHAPTER 1: PHOTONS AND GRAVITONS—SCRAPPED

Early in the 20th century, Einstein and others found that mass and energy are merely two ways of describing the same thing ($E = mc^2$). That is, a particle which has energy must have mass, and conversely.

About the same time, Einstein developed the concept of a wave packet (later named a photon). This was his explanation of how light is transmitted. But this concept violates the idea behind $E = mc^2$, since it presumes that the photon has energy but no mass (else how could it travel at the speed of light?).

Since then, there have been many physical demonstrations that $E = mc^2$ is correct. In contrast, there have been no such demonstrations showing that photons exist.

Of course, light does get transmitted. If it is not transmitted by photons, then how is it transmitted?

Somewhat later, the concept of graviton was introduced to explain how gravity works. The concept parallels that of the photon, and has similar problems.

SECTION 1A: CONNECTIOON THEORY

Connection theory is the theory of how particles interact. The basic idea behind the theory is that two particles interact when in some sense they become close enough to each other to form a *connection*. The basic details of the theory are given in this section.

Obviously particles communicate with one another, both electrically and gravitationally. If not by means of photons or gravitons, then by what means do such communications take place?

A clue is found in the form of primitive particles (that is, electrons, positrons, protons, and antiprotons). The clue is that each of these particles has a frequency. In fact, the frequency of an electron or positron at rest is 1.24×10^{20} cycles per second, and that of a proton or antiproton at rest is 2.27×10^{23} cycles per second. And by special relativity, the frequency increases if the particle is moving.

Having a frequency means that these primitive particles vibrate! So what is going on to produce these vibrations?

Consider the electron. The energy E of an electron at rest is related to its frequency ν by Planck's constant $h = 6.626\,070 \times 10^{-34}$ joule-seconds:

$$E = h\nu$$

This energy is constant. No matter how long the electron exists at rest, it does not gain or lose any energy, unless it enters into an interaction with one or more other particles.

So how does this limitless supply of energy relate to the frequency of the electron?

ASSERTION: The number of probes emitted per second by a primitive particle is equal to the frequency of that particle.

Many details are involved in adequately describing a probe. Some will be presented here; others will be shown later.

To proceed, it is convenient to define the *core* of an electron. The core is a placeholder, the center of mass of the electron.

The probe is emitted from the core of the electron. In one frequency cycle, the probe flies out in a straight line to some great distance, and then returns to its core. The next cycle begins immediately, and the probe is emitted in almost the opposite direction. The difference between the opposite direction and *almost* the opposite direction is the cause of particle spin.

The ideas of the preceding paragraphs apply also to positrons, protons, and antiprotons, using the appropriate frequencies.

On its outward journey, a probe (from any primitive particle) may come close enough to the core of another primitive particle to make a probe-core connection with that particle. It may also come close enough to the probe of another primitive particle to make a probe-probe connection with that probe. A connection allows for an exchange of energy between the probe and the core, or between the two probes.

More information about connection theory is scattered throughout the rest of this book.

Evidence from the deuteron (the nucleus of the deuterium atom) yields several properties of probe behavior in such connections.

SECTION 1B: A DEUTERON DETOUR

The deuteron is the nucleus of the deuterium atom, and consists of a proton plus a neutron. A neutron consists of a close combination of a proton and an electron. Thus the deuteron consists of a close combination of two protons and one electron.

As described above, primitive particles vibrate Thus the three particles of the deuteron must emit at least as many probes as they would if all three were at rest. Corresponding values are shown in the next table.

Number of Probes Emitted per Second by Particles at Rest

Particle	Frequency $\left(s^{-1}\right)$
Electron	$1.235\,559 \times 10^{20}$
Proton	$2.268\,732 \times 10^{23}$

From this table, it follows that the three particles of the deuteron must emit at least this many probes per second:

$$1.235\,559 \times 10^{20} + 2 \times 2.268\,732 \times 10^{23} = 4.538\,700 \times 10^{23} \equiv P_1.$$

This corresponds to a mass of $3.346\,155 \times 10^{-27}$ kg. (This follows from the special relativity formula $mc^2 = h\nu$, where m is the particle mass, c is the velocity of light in a vacuum, h is Planck's constant, and ν is the particle frequency.) And it is quite likely that these nuclear constituents are moving, so that their frequencies are greater than the rest frequencies.

However, the measured mass of the deuteron is only $3.343\,583 \times 10^{-27}$ kg., which corresponds to $4.535\,211 \times 10^{23} \equiv P_2$ probes per second. Thus the deuteron would seem to have a deficiency of at least $P_1 - P_2 = 3.489 \times 10^{20}$ probes per second. How can this deficiency be explained?

Let N_E be the number of probes emitted per second by the particles in the deuteron. Suppose that N_A of them are absorbed by one of the other nuclear particles, and that the remainder, $N_M = N_E - N_A$, miss those other particles. Then the N_M probes which miss are the only ones that can be observed, and so they account for the measured mass of the deuteron. Thus

$$N_M = P_2 = 4.535\,211 \times 10^{23}$$
$$N_E \geq P_1 = 4.538\,700 \times 10^{23},$$

hence $N_A = N_E - 4.535\,211 \times 10^{23} \geq 3.489 \times 10^{20}$.

It seems reasonable to assume that the N_A absorbed probes account for the binding energy of the deuteron, which is about 2.224 52 MeV. This is equivalent to $3.965\,566\,734 \times 10^{-30}$ kilograms, and to a frequency of $5.378\,865\,406 \times 10^{20}$ probes per second. That is, $N_A = 5.378\,865\,406 \times 10^{20}$.

This implies that sometimes a probe emitted by a nuclear particle will connect with another particle in the same nucleus and will not proceed outside of the nucleus. That is, when a probe hits another particle, it stops there and does not continue on its outward journey.

A further implication is that the measured mass of an atom is based on the probes which emerge from the atom, but does not count the probes which connect with other particles in the atom (either in the nucleus or among the electrons orbiting the nucleus).

SECTION 1C: THE STRONG NUCLEAR FORCE

Every atom with a nucleus larger than the deuteron will have the same feature as deuterium, namely, that occasionally the probe of a nuclear constituent will hit the core of such constituent, and on that cycle will not be able to proceed outside the nucleus.

The energy produced by such nuclear probe-core connections within an atom is the source of the strong nuclear force.

SECTION 1D: RESULTS OF A CONNECTION

The result of a probe-probe connection is first, that each probe changes direction slightly and continues on its cycle. This is implied by the observed bending of light beams passing stars.

The second result of a probe-probe connection is that each of the corresponding cores receives a slight impulse. Each such impulse will either be toward or away from the other core. The accumulation of these slight impulses is the cause of the gravity between the two cores.

There is no limit to the number of probe-probe connections for a probe on a single cycle. This is implied by the fact that light is observed on Earth after having passed very close to a star

The first result of a probe-core connection is an exchange of energy between the probe and the core. In particular, if the source particle of the probe had acquired extra energy from a previous probe-core connection, that extra energy is carried by the probe and in the present connection, may be transferred to the target core. This is how light is transferred from one particle to another. And in addition, this is how electromagnetic interactions occur between particles.

The second result of a probe-core connection is an end to that cycle of the probe. That is, instead of proceeding, it returns to its core and begins another cycle.

CHAPTER 2: NEWTON'S LAW OF GRAVITY—CORRECTION 1

Newton's law of gravity:

$$F = \frac{m_1 m_2}{r^2},$$

approximates the force of gravity F between two masses m_1 and m_2 that are separated by a distance r. This equation seems to imply that as the distance $r \to 0$, the force $F \to \infty$. This conclusion is not correct. A correct statement is that the only values of r for which Newton's law is valid lie in the range $0 < r_{g1} \le r \le r_2 < \infty$, where r_{g1} is a small number (perhaps on the order of 1 or less), and r_2 is roughly 20 billion light years (the maximum distance that a probe can go, hence the maximum distance that light can go).

No experimental evidence regarding the value of r_{g1} seems to exist. However, the fact that $r_{g1} > 0$ will be demonstrated below.

The estimate that r_2 is about 20 billion light years is based on a study of 84 stars by Allan Sandage.

Now consider two primitive particles separated by a distance r. Electrical interactions between the two particles occur when the probe of either particle connects with the core of the other particle. Gravitational interactions between the two particles occur when the probe of one particle connects with the probe of the other particle.

One deduction can be made immediately: Gravity is not propagated by fields. Rather, it is propagated by finite sequences of probe-probe connections.

What happens to the gravity between these two particles as $r \to 0$? If the two particles remain at or near rest, then the number of probes emitted per second remains finite for each particle, and hence the probability of a probe-probe connection between them remains finite, as does the gravity between them. If the particles gain energy, then of course the probability of probe-probe connections increases, as does the gravity between them.

But what would cause the energy of the two particles to increase? Up to a certain point, the energy increases simply because as the two particles get nearer to each other, the probability of a probe-probe connection increases. After a certain minimal separation is reached, however, a smaller separation yields no increase in the number of probe-probe connections. Nothing happens between the two particles to increase the energy of either particle.

Thus the energy of the pair of particles remains finite, and does not approach infinity. And thus Newton's formula does not provide a correct description of the situation as $r \to 0$.

Alternatively, an infinite value of gravitational force would require an infinite number of probe-probe hits in a finite amount of time. This would require that each particle must obtain an infinite apparent mass in a finite amount of time, which does not seem possible.

Hence it seems that no matter how small r becomes, the gravity between the two particles remains finite, and so Newton's law no longer describes the situation for very small r.

SECTION 2A: HOW GRAVITY WORKS

In the case of two probes connecting, there are two locations where such connections are far more likely to occur than elsewhere. These locations are the volumes closely surrounding each of the probe sources. The reason is as follows. Consider a sphere of constant volume located somewhere in space. The number of times that the probe from a given particle passes through this sphere depends on where the sphere is located. For probes from particle 1, that number is maximized if the center of the sphere is located at the center of mass of particle 1. The further the center of the sphere is from particle 1, the fewer the number of times the probe from that particle will pass through the sphere. A similar statement applies to probes from particle 2.

Thus the volume in which probes from particle 1 are most dense is centered on particle 1, and that for which probes from particle 2 are most dense is centered on particle 2.

When two probes connect, there is an exchange of energy between the two probes. This results in two attractive forces, one between the point of contact and the first source particle, the other between the point of contact and the second particle. The possible consequences are shown in the following graphs. We will explore only the gravitational consequences for particle 2, but of course there is an analogous effect on particle 1.

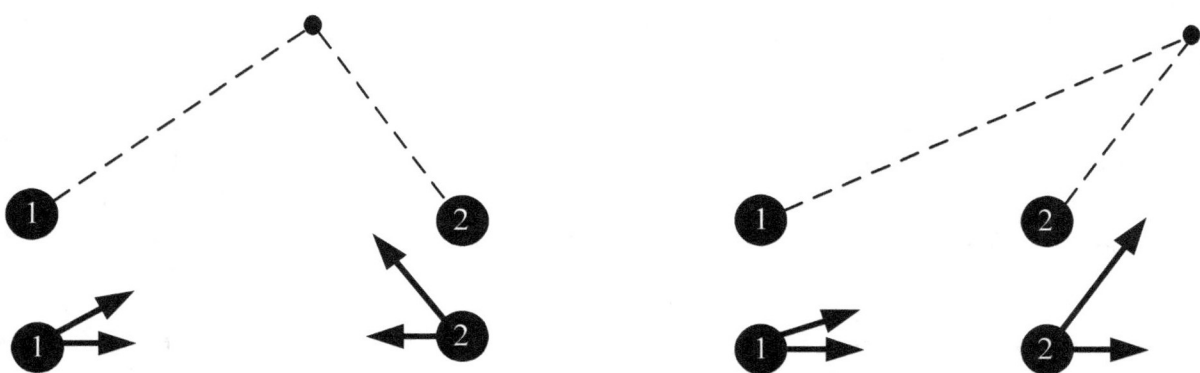

The upper left graph shows the case that the two probes meet at a point on the left side of particle 2. This leads to a tiny attractive force between each particle and the point of contact. Those forces are shown in the lower left graph, along with the resulting forces in the direction between the two particles. In this case, particle 2 is pulled slightly toward particle 1.

8

The upper right graph shows the case that the two probes meet at a point on the right side of particle 2. This leads to a tiny attractive force between each particle ant the point of contact. These forces are shown in the lower right graph above, along with the resulting forces in the direction between the two particles. In this case, particle 2 is pushed slightly away from particle 1.

It is apparent that there are likely to be about the same number of connections that push particle 2 away from particle 1 as there are that pull particle 2 toward particle 1. However, there is another factor that enters into this analysis, and that is that every particle has a hit-free zone. In these cases, the hit-free zone of particle 2 is at work. It was stated above that when a probe hits a core, then it cannot go further on that cycle, but instead returns to its own core.

Because of this property of probes, every particle has a hit-free zone, as illustrated in the next graph.

The idea is that any probe from particle 1 which is aimed at a point in the hit-free zone of particle 2 will hit particle 2 itself before it can enter the hit-free zone. And once it hits particle 2, it can no longer proceed, but returns to its own core.

One would presume that it is equally likely for probes from particle 1 to go in any direction. Hence it would seem that such probes would hit just as many probes from particle 2 on the left side of that particle as on the right side of it. But the existence of the hit-free zone results in a smaller number of hits on the right side than on the left side. Now the energy generated by any probe-probe hit will be roughly the same. Hence the total energy of those hits on the left side of particle 2 will be slightly more than that on the right side. Thus particle 2, on the average, is pulled slightly toward particle 1. And this is how gravity works.

After the hit, the two probes will then continue their respective cycles, but in slightly different directions.

Such probe-probe hits are the sole cause of gravity.

CHAPTER 3: COULOMB'S LAW—CORRECTED

Coulomb's law:

$$|F| = k \frac{|q_1 q_2|}{r^2}$$

yields the magnitude $|F|$ of the electromagnetic force between two particles with charges q_1 and q_2 separated by a distance r, where k is a constant. This equation seems to imply that as the distance $r \to 0$, the force magnitude $|F| \to \infty$. This conclusion is not correct. A correct statement is that the only values of r for which Coulomb's law is valid lie in the range $0 < r_{el} \le r \le r_2 < \infty$, where r_{el} is a small number (perhaps on the order of 1 or less), and r_2 is roughly 20 billion light years (the maximum distance that a probe can go, hence the maximum distance that light can go).

Electromagnetic force arises when the probe of a particle P_1 connects with the core of a second particle P_2. Clearly the chance of this happening depends on the distance between the two particles. For distances r greater than particle size, the probability of a connection is

$$P = \frac{k_e}{4\pi r^2},$$

where k_e is a constant, and the denominator is the surface area of a sphere of radius r.

But when the two particles get close enough to each other, the probe from particle 1 will connect with the core of particle 2 on every other cycle of particle 1. (It can't connect every time, because the particles are apart from each other, and probe 1 goes out in almost opposite directions on successive cycles.) Hence in this case, the probability is simply

$$P = \frac{1}{2}.$$

This probability holds until the two particles are so close that they touch. At that point, a core-core connection arises, the energy of which vastly exceeds that of a probe-core connection.

In conclusion, the probability of a probe-core connection never exceeds ½. Thus the energy of such a connection remains finite for all values of r.

CHAPTER 4: BLACK HOLES— SCRAPPED

According to Wikipedia,

(1) A black hole is a region of spacetime exhibiting such strong gravitational effects that nothing—not even particles and electromagnetic_radiation such as light—can escape from inside it.

(2) The theory of general relativity predicts that a sufficiently compact mass can deform spacetime to form a black hole.

Paragraph (1) above is based on an erroneous concept of light, as well as an erroneous concept of gravity. Light is regarded as being propagated by photons. Gravity is regarded as being propagated by gravitons, having no limits, and having the ability to stop the movement of light.

As explained in Chapter 1, this book scraps the idea of photons and gravitons. Instead, gravity results from connections between two probes.

Such a connection results in four things: a tiny impulse on each of the cores which emitted the probes, and a tiny change in the direction of each probe on the remainder of its current cycle. The impulses on the two cores can have four results: core 1 may be either attracted to or repulsed from core 2, and core 2 may be either attracted to or repulsed from core 1. These are the ONLY effects of gravity. Thus a single such connection has NO direct effect on any other particle

The surface of a star, no matter how massive the star, consists of particles, some of which are electrons and some of which are protons, possibly part of an atom. Regardless of the local gravity, an electron or a proton must still vibrate. And some of the probes from these primitive particles will head away from the star. These probes will often carry extra mass (potential light).

It follows from previous chapters that the only thing that can stop a probe is when it connects with another primitive particle. But outwards from the surface of the star, there are very few primitive particles to be found. Hence these probes will keep going away from the star, and consequently will emit light to be seen by Earth! Thus no star can be a black hole.

Paragraph (2) above is a conclusion drawn from a theory which is only an approximation to reality, as will be shown in a later chapter. This conclusion assumes that a "sufficiently compact mass" can exist. However, in Chapter 2 above, it was shown that a star cannot shrink to zero under infinite gravity. And there is not a shred of physical evidence to indicate that a star can somehow be squeezed into some sort of "compact mass". So it seems inescapable to conclude that the prediction of paragraph (2) is false. In other words, a star cannot be a black hole.

CHAPTER 5: EXPANDING UNIVERSE AND BIG BANG— SCRAPPED

Einstein's 1916 theory of general relativity seemed to imply that the universe was expanding, but this was not taken seriously until Hubble advanced his own theory in 1927.

Georges Lemaître had published a theory of an expanding universe in 1925, but it was in a Belgian journal and received little notice until about 1931, when it was translated into English by Arthur Eddington.

So Edwin Hubble became recognized as the chief proponent of an expanding universe.

In about 1929, Fritz Zwicky noted that light seemed to lose energy as it passed through space (tired light). But he could not offer an adequate explanation for this effect.

However, it was shown above that probes are the means by which particles communicate with other particles. But what are probes made of?

It turns out that a probe actually consists of the mass/energy of the particle, which sort of unrolls, leaving behind a tiny trail of mass/energy, until it runs out of mass/energy. It then sort of rolls back up that tiny trail, arriving back at the particle's core with full mass/energy restored. The next probe is then emitted in almost the opposite direction. (I regret the vagueness of this description, but better words do not come to mind.)

In consequence, the mass/energy of the probe tip diminishes with distance. Hence if it happens to connect with another core, the amount of mass/energy available for the connection to the other core is also diminished.

More explicitly, say the intrinsic mass of the particle is m. After the probe has gone a distance D, the intrinsic mass at the probe tip has diminished to $m(1 - D/R)$, where R is the maximum distance that the probe can go. This is then the mass brought to bear in the connection with the target particle.

Note that the mass m in the preceding paragraph covers the cases that the source particle is either at rest or not, and also whether the source particle contains added energy or not. In the case that the source particle contains added energy this added energy may be deposited on the target particle during the connection. This deposit being from a probe tip with reduced energy $m(1 - D/R)$ is the cause of light from a distant star appearing to have less than full energy (the Zwicky effect).

Traditionally, the redshift of light from a star is calculated as follows:

$$z = \frac{\lambda_{rec} - \lambda_{em}}{\lambda_{em}},$$

where λ_{em} is the emitted wavelength and λ_{rec} is the received wavelength. This value presumably shows the difference in velocities between a star which is stationary with respect to Earth, and one which is moving away from Earth.

However, if the star is a distance D from Earth, then the energy of the light diminishes by a factor of $1 - D/R$. Thus for a star which is stationary with respect to Earth, the light arriving from that star will have a frequency of

$$\lambda_0 = (1 - D/R)\lambda_{em}.$$

Consequently, the correct measure of the velocity of the star with respect to Earth is the difference between λ_0 and λ_{rec}, and the correct redshift is

$$z = \frac{\lambda_{rec} - \lambda_0}{\lambda_0}.$$

ASSERTION: Applying this correction to existing star data will result in an average value of 0 for the radial velocities of stars with respect to Earth.

This assertion has been shown to be correct for 60 stars in a set of 84 stars listed by Allan Sandage. The nonzero values of the other 24 stars merely says that there are many stars which move radially away from or towards Earth.

Clearly, if this assertion can be shown to be true for a much larger set of stars, then the idea of an expanding universe will be shown to be false, and hence the idea of a Big Bang is equally false.

CHAPTER 6: GENERAL RELATIVITY—CORRECTED

In its orbit around the Sun, a planet's closest approach to the Sun is called its perihelion. Over time, the location of this perihelion changes; this is called the precession of the perihelion. For a long time after Newton presented his law of gravity (1686), it seemed that this law would explain the precession of the perihelion of every planet.

In 1859, the French mathematician and astronomer Urbain Le Verrier, using the latest astronomical measurements and very accurate calculations, discovered a discrepancy between the predictions of Newton's law of gravity and the observed precession of the perihelion of the planet Mercury. He decided that this discrepancy must be due to an error in Newton's law of gravity.

For the next 57 years, astronomers and mathematicians struggled without success to explain this discrepancy.

In 1916, Einstein presented his theory of general relativity. One of the claims made for the validity of the theory was that it allowed the calculation of the values of the perihelion precessions of the planets in the Solar System. Thus for the next century it was believed that the explanation of this discrepancy was general relativity. However, this explanation seemed a bit weak, since it did not offer any reason for the corresponding movements of the planets.

Early in 2014, I was examining those calculated values. After pondering these values for a while, it seemed to me that they could be explained if the planets were not receiving the total gravitational pull of the Sun. Specifically, it appeared to me that instead of orbits based on the center of mass of the Sun, the orbits of the planets were based on an offset center of mass of the Sun (a *gravitational offset*).

On the next page is an illustration of a gravitational offset in a star (not to scale!). The gravitational offset equals the distance between the true center of mass and the apparent center of mass. The volume of a star used in determining the apparent center of mass consists of all the points to the right of the true center of mass, and most but not all of the points to its left. These are the points within the star (or Sun) that can form a gravitational attraction with a planet on the right side of the star. Some of the points on the left side of the star cannot reach a planet to the right side of the star. This gravitational blockage is what causes the gravitational offset to exist. (An explanation of gravitational blockage will be presented below.)

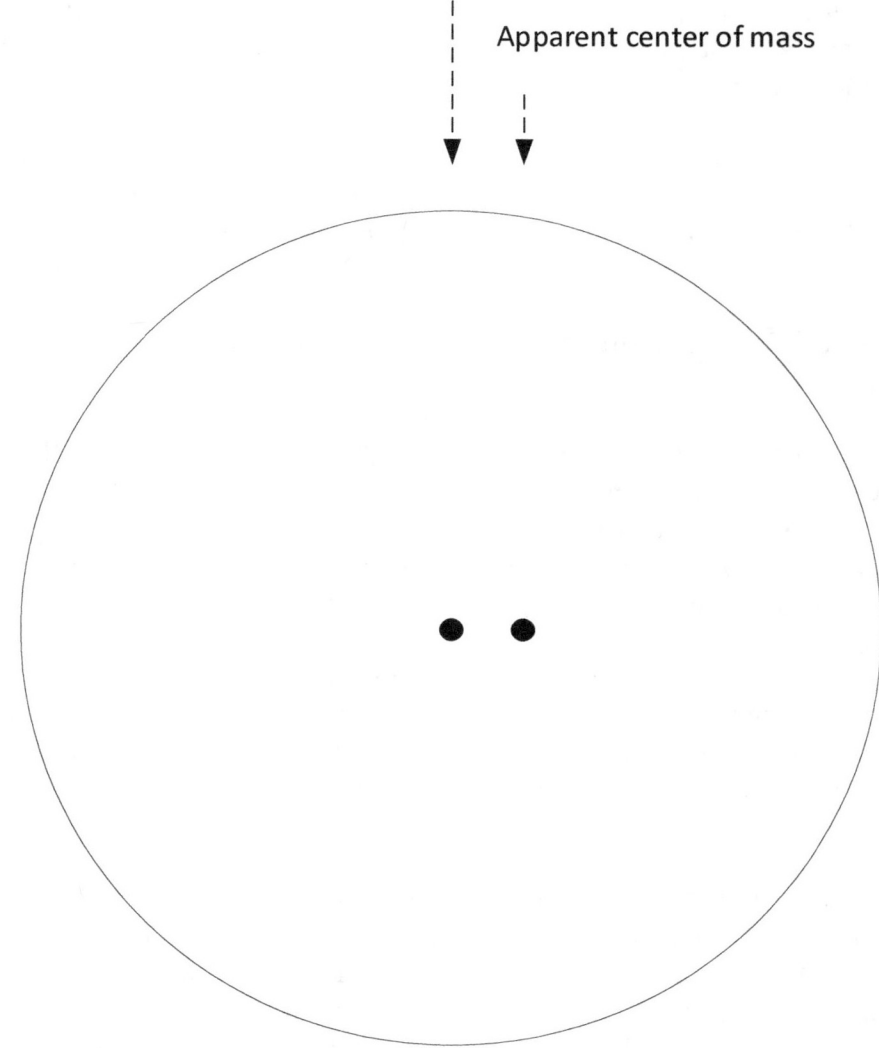

True center of mass

Apparent center of mass

I used classical celestial mechanics (F. R. Moulton, An Introduction to Celestial Mechanics, 1914) to develop the equations for a planet orbiting the Sun in the case that the Sun does have a gravitational offset. For details, see the reference *General Relativity: Not Exact, but a Useful Approximation.*

One result of these equations is that they provide a simple formula relating the Sun's gravitational offset and the planet's perihelion precession.

A second result is that the resulting equations yield very accurate values for the planetary orbits, as well as very accurate values of the perihelion precessions, all using Newton's law of gravity.

Thus Le Verrier was wrong: Newton's law of gravity works correctly in this situation. Le Verrier simply did not know how to *use* it correctly. Here is Newton's formula, with a corrected explanation for its use. The gravitational force F between the Sun and a planet is

$$F = \frac{Gm_S m_p}{r^2},$$

where G is Newton's gravitational constant, m_S is the mass of the Sun, m_p is the mass of the planet, and r is the distance between the *apparent* center of mass of the Sun and the center of mass of the planet.

Thus the correction proposed for general relativity is simply to say that it is an excellent approximation for things such as planetary perihelia precessions, but it is not an exact theory. The existence of gravitational offsets provides more exact values for these precessions

Another limitation of general relativity is that in the case of planetary precessions, it does not seem to provide any physical explanation for the corresponding planetary movements. On the other hand, the existence of gravitational offsets provides a straightforward explanations (via Newton's law of gravity) for the corresponding movements of the planets.

Hence I conclude that general relativity is a nice approximation, but not an exact description of physical situations. Consequently, it follows that theories based upon the assumptions behind general relativity may well be erroneous.

SECTION 6A: A PATH THROUGH THE SUN

Physical evidence (the precessions of planetary perihelia) indicates that the Sun possesses a gravitational offset. But what causes it?

Consider a primitive particle on one edge of the Sun. Suppose it emits a probe in the direction of the center of the Sun. On its path through the Sun. it will pass near a vast number of cores of other particles. The existence of a gravitational offset in the Sun suggests that the path of this probe is almost certain to make a connection with the core of another particle in the Sun, which will stop the probe. Hence the probe never reaches the other side of the Sun. This phenomenon is the cause of the gravitational offset.

Any star at least as massive as the Sun will have a gravitational offset. This is obvious for ordinary stars. For dwarf stars, this is shown to be true by the behavior of the binary pulsar B1913+16. For details, see the paper by J. M. Weisberg and J. H. Taylor, Relativistic Binary Pulsar B1913+16: Thirty Years of Observations and Analysis, http://arxiv.org/PS_cache/astro-ph/pdf/0407/0407149v1.pdf.

CHAPTER 7: NEWTON'S LAW OF GRAVITY—CORRECTION 2

In view of the previous chapter, it is simple to adapt Newton's law of gravity to the case that a star (such as the Sun) has a gravitational offset. Newton's law, when applied to two celestial bodies, is correct when the usual formula:

$$F = \frac{Gm_1m_2}{r^2},$$

is interpreted in the following way:

F is the attractive force between the apparent centers of mass of each body,

G is Newton's gravitational constant,

m_1 is the gravitational mass of the first body,

m_2 is the gravitational mass of the second body,

r is the distance between the true centers of mass of the two bodies, minus the sum of the gravitational offsets of the two bodies.

PART 2: A SUMMARY OF CONNECTION THEORY

CHAPTER 8: THE UNIVERSE

The universe contains a very large number of primitive particles (electrons, positrons, protons, and antiprotons). Whether this number is finite or infinite is something that humans are unlikely ever to discover.

These primitive particles are all in constant motion, resulting in the transfer of vast amounts of energy from particles to other particles. Each quantum of this energy is physically a part of one particle until it is transferred to another particle. Energy never exists apart from a particle (core plus probe; see below).

Every primitive particle has two parts, called its core (placeholder) and its probe. Probes are emitted from the core. The number of probes emitted per second by the particle is equal to the frequency of the particle. A cycle is the time between the emission of a probe and its return to its core.

On its outward journey, a probe moves in a straight line from the core. If it hits another probe, its direction is changed slightly. It continues in this way until either it hits another core or it runs out of *em* (an abbreviation for energy/mass). Then it returns to its core, thus ending a cycle. The next cycle begins immediately, and the probe goes out in almost the opposite direction. That 'almost' is the cause of particle spin.

A probe is emitted with nearly the entire em of the particle; a tiny bit of em is left at the core. As the probe moves through space, it leaves behind it a tiny trail of em. When it is stopped (either because it hits a core or runs out of em), it then returns to its core, gathering all the em it had left behind. On its return journey, it cannot connect with any other core or probe.

If, on its outward journey, a probe hits the core of another particle, its contribution to the em balance of the hit is its current amount of em, not the original em of the particle. Similarly, the contribution of the other particle to the em balance is whatever remains of the em of that particle at its core, assuming that its probe is off somewhere else.

There are three types of hits that can occur. Two of them are suggested by the above words: a probe-core hit, and a probe-probe hit. The former type is responsible for electromagnetic energy transfers, and the latter is responsible for gravity. The third type of hit is a core-core hit. The meager evidence available suggests that such hits are rare, and last for only a very brief time.

If a particle has engaged in a previous probe-core hit, it may have an increased amount of em. If so, then on its next cycle, its probe will carry that extra em, and if a probe-core hit occurs, may deposit it on the target core. This is how light is transferred from one particle to another.

An atom is held together by the probe-core hits between the particles in the nucleus and the particles in orbit about the nucleus. The charge of these particles plays no part in keeping the particles together.

This summary of connection theory is probably incomplete. Users are invited to suggest other likely properties of the theory.

REFERENCES

The following books by the present author contain details of much of the material in this book, and are available at Amazon.com:

Gravity: What It Is and How It Works, July 27, 2016.
General Relativity: Not Exact, But a Useful Approximation, December 17, 2016.
The Physical Basis of Tired Light, Revised Edition, March 20, 2017.
A Direct Contact Approach to Physics, July 19, 2017.